Marek Berezowski

Fractals gallery of complex numbers

Marek Berezowski

Fractals gallery of complex numbers

LAP LAMBERT Academic Publishing

Impressum / Imprint

Bibliografische Information der Deutschen Nationalbibliothek: Die Deutsche Nationalbibliothek verzeichnet diese Publikation in der Deutschen Nationalbibliografie; detaillierte bibliografische Daten sind im Internet über http://dnb.d-nb.de abrufbar.

Alle in diesem Buch genannten Marken und Produktnamen unterliegen warenzeichen-, marken- oder patentrechtlichem Schutz bzw. sind Warenzeichen oder eingetragene Warenzeichen der jeweiligen Inhaber. Die Wiedergabe von Marken, Produktnamen, Gebrauchsnamen, Handelsnamen, Warenbezeichnungen u.s.w. in diesem Werk berechtigt auch ohne besondere Kennzeichnung nicht zu der Annahme, dass solche Namen im Sinne der Warenzeichen- und Markenschutzgesetzgebung als frei zu betrachten wären und daher von jedermann benutzt werden dürften.

Bibliographic information published by the Deutsche Nationalbibliothek: The Deutsche Nationalbibliothek lists this publication in the Deutsche Nationalbibliografie; detailed bibliographic data are available in the Internet at http://dnb.d-nb.de.

Any brand names and product names mentioned in this book are subject to trademark, brand or patent protection and are trademarks or registered trademarks of their respective holders. The use of brand names, product names, common names, trade names, product descriptions etc. even without a particular marking in this works is in no way to be construed to mean that such names may be regarded as unrestricted in respect of trademark and brand protection legislation and could thus be used by anyone.

Coverbild / Cover image: www.ingimage.com

Verlag / Publisher:
LAP LAMBERT Academic Publishing
ist ein Imprint der / is a trademark of
OmniScriptum GmbH & Co. KG
Heinrich-Böcking-Str. 6-8, 66121 Saarbrücken, Deutschland / Germany
Email: info@lap-publishing.com

Herstellung: siehe letzte Seite /
Printed at: see last page
ISBN: 978-3-659-63075-0

Copyright © 2014 OmniScriptum GmbH & Co. KG
Alle Rechte vorbehalten. / All rights reserved. Saarbrücken 2014

Fractals gallery of complex numbers
Black and white fractals power

Marek Berezowski

For all

Table of contents

Part 1. Introduction 5
Part 2. The complex squared hyperbole 6
Part 3. The power complex 17
Part 4. Chemical reactor models 40
References 54

Part 1. Introduction.

This book presents a set of fractals, which were created as a visualization of the scientific results of the author on the nonlinear dynamics. This collection has been divided into three main parts due to the three different mathematical models. Complex numbers were used to describe the models.

The first part refers to a model based on the complex hyperbole raised to the square. The second part refers to a model in which the complex variable has been raised to a power given by complex variable. The third part concerns mathematical models of the chemical reactor. This model consists of partial differential equations with appropiated boundary conditions. All three above mentioned models are presented at the beginning of each section of this work in detail.

The complete form of the fractal was first presented in each part. Subsequent figures are fragments thereof. The form and shape of these fragments are similar to each other which proves their fractal structure.

Fractal images created by the autor have not only aesthetic value. They also allow the evaluation of the sensitivity and stability of the model. Lyapunov exponent and shade scattering proved about this. In the original, these fractals are colourful [1].
Because of appearance of some fractals, they were given names.

Part 2. The complex squared hyperbole

The fractals on Figs. 2.1 to 2.10 were generated from the following requrence:

$$z \to \frac{1}{z^2} + c = F(z,c) \qquad (2.1)$$

where z and c are complex numbers. The condition of convergence of the above sequence is:

$$\left|\frac{dF}{dz_s}\right| \leq 1 \Rightarrow |z_s| \leq \sqrt[3]{2} \qquad (2.2)$$

where

$$z_s - \frac{1}{z_s^2} = c \qquad (2.3)$$

what gives the limit:

$$c = \sqrt[3]{2}e^{i\varphi} - \frac{1}{\sqrt[3]{4}}e^{-2\varphi}. \qquad (2.4)$$

The line which is the contour of the fractal on Fig.2.1 is abtained by changing the value of parameter φ from 0 to 2π. The shade of fractals on Figs.2.1 to 2.4 correspond to the values of $|z|$.

The Lyapunov exponent is the parameter, which indicates the stability of sequence obtained from (2.1). It is determined by the formula:

$$\lambda = \lim_{n \to \infty} \frac{1}{n} \sum_{i=1}^{n} \ln\left|\frac{dF}{dz_i}\right|. \qquad (2.5)$$

For the requrence (2.1) λ has the form:

$$\lambda = \lim_{n \to \infty} \frac{1}{n} \sum_{i=1}^{n} \ln\left(\frac{2}{|z_i^3|}\right). \qquad (2.6)$$

The shade of fractals on Figs.2.5 to 2.10 correspond to the values of λ.

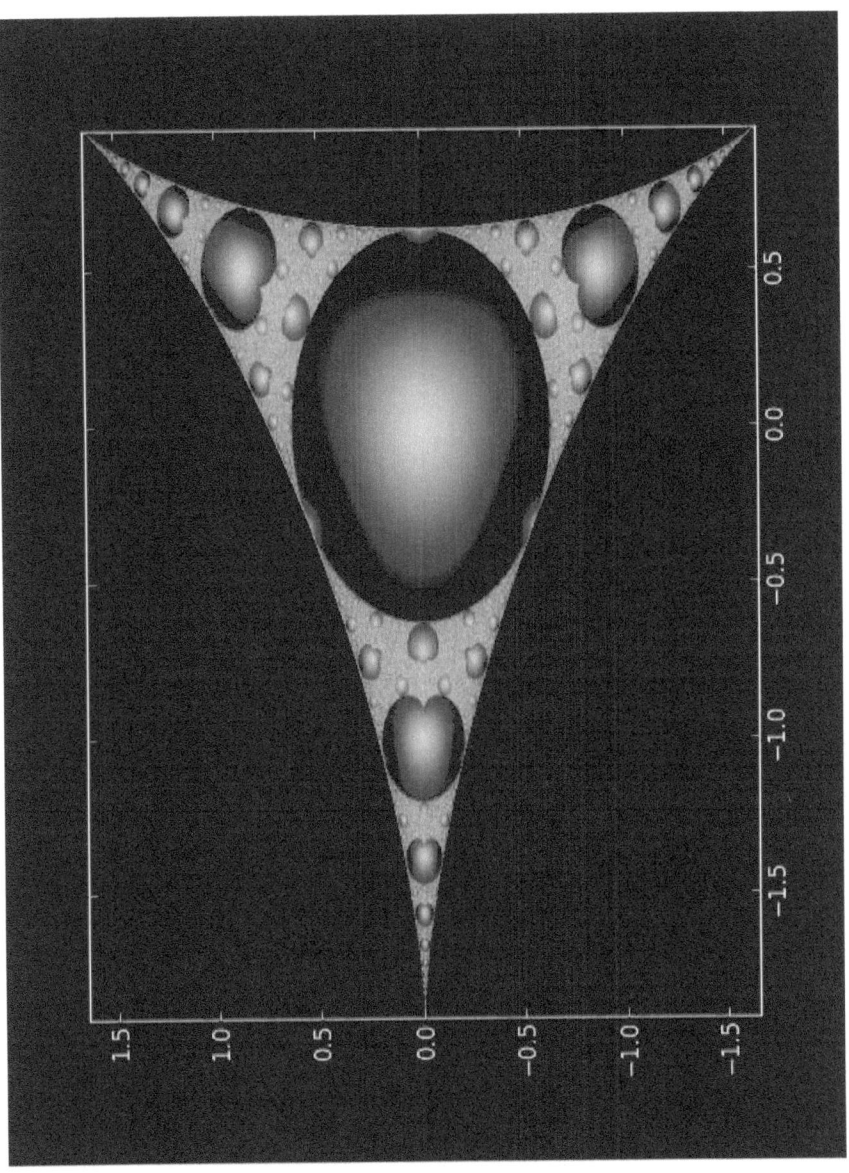

Fig. 2.1. Full set.

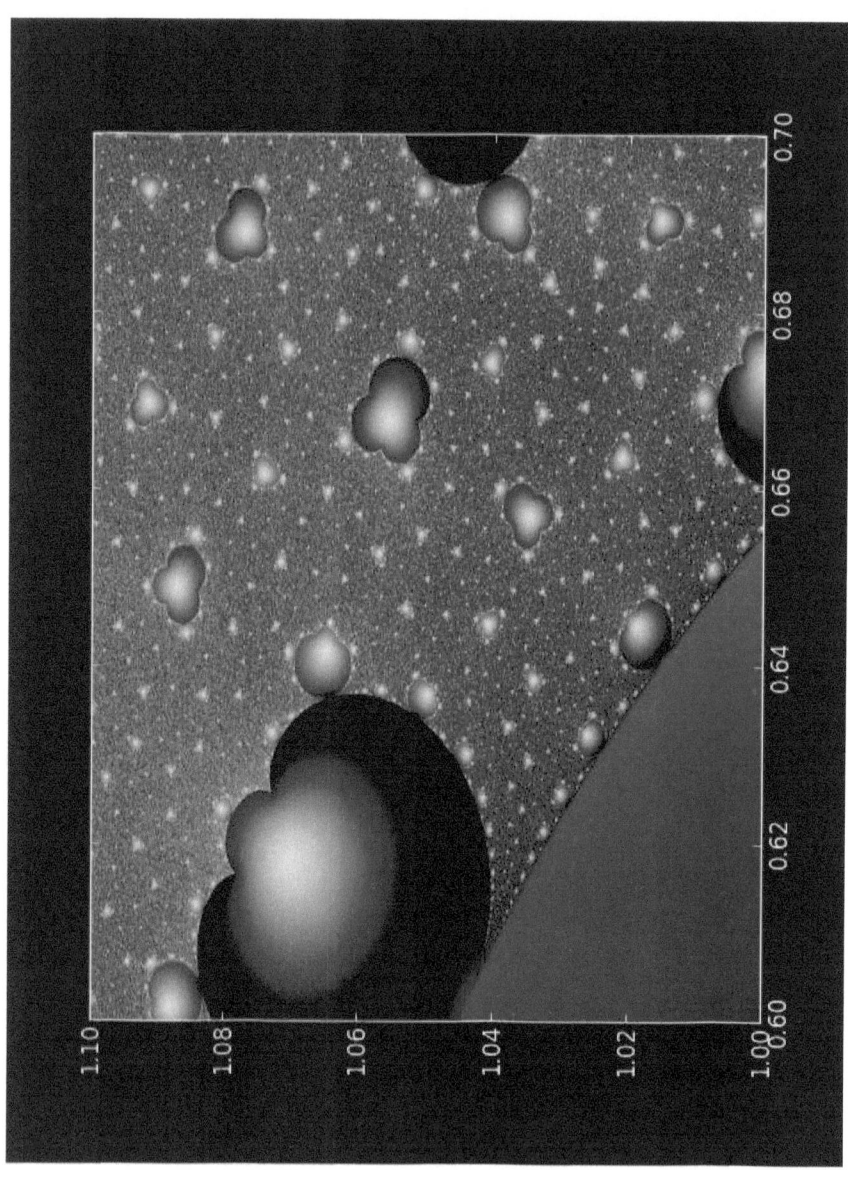

Fig. 2.2. Drops.

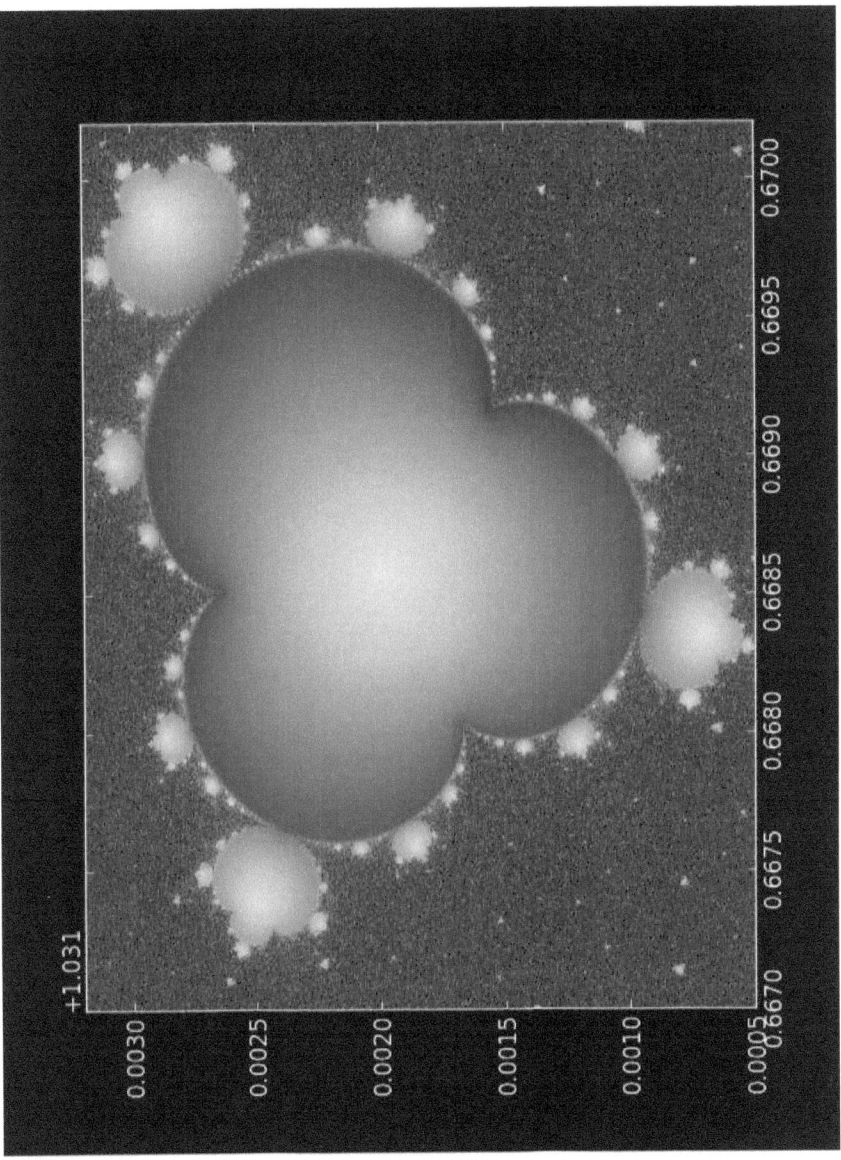

Fig. 2.3. Similar to Mandelbrot's fractal.

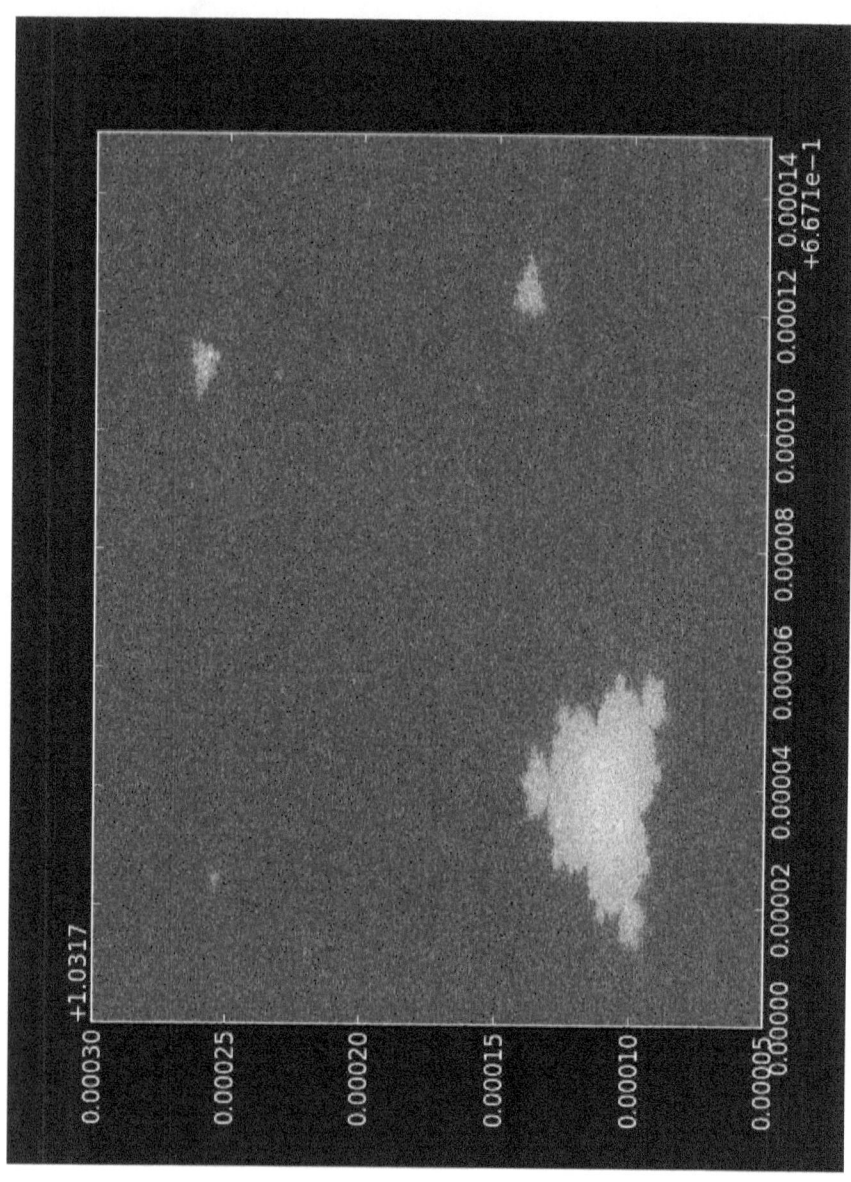

Fig. 2.4. Ufo.

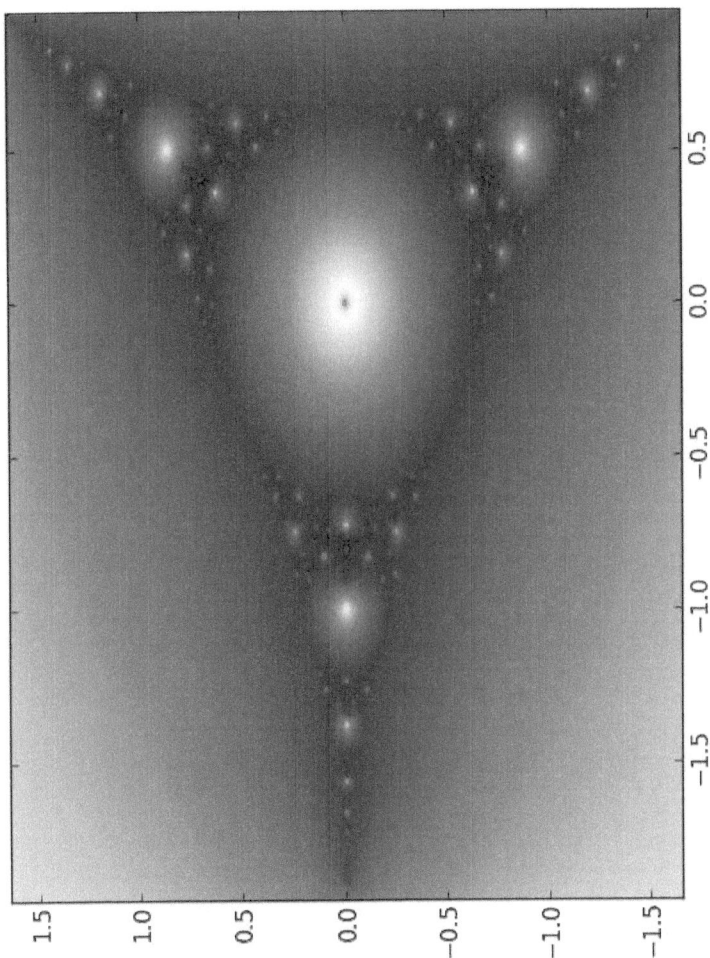

Fig. 2.5. Lyapunov exponent - full set.

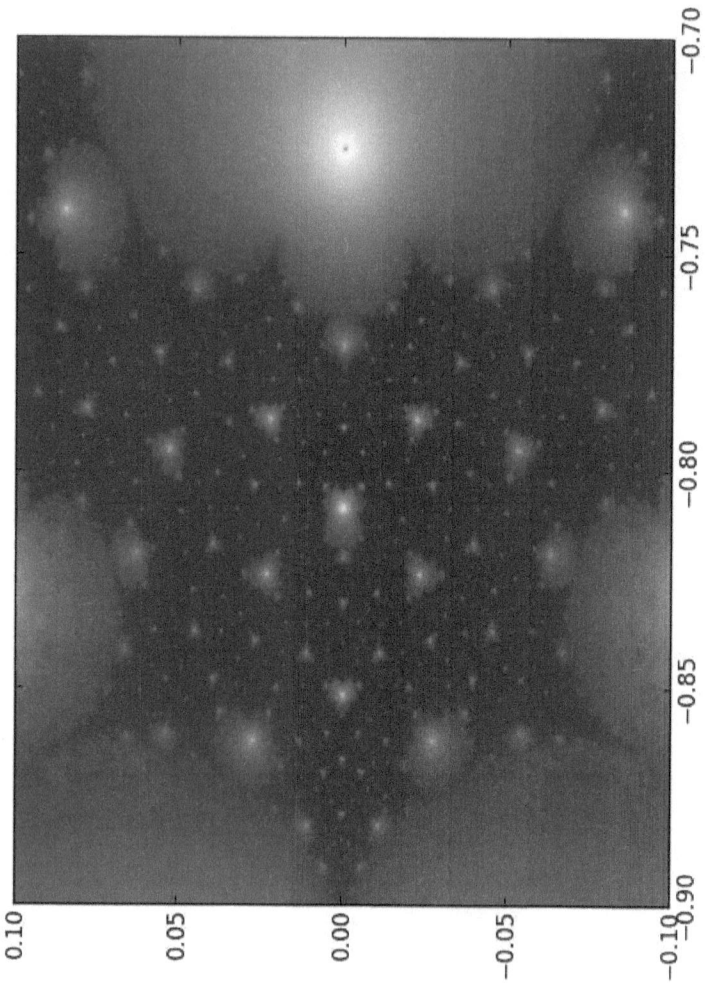

Fig. 2.6. Invasion.

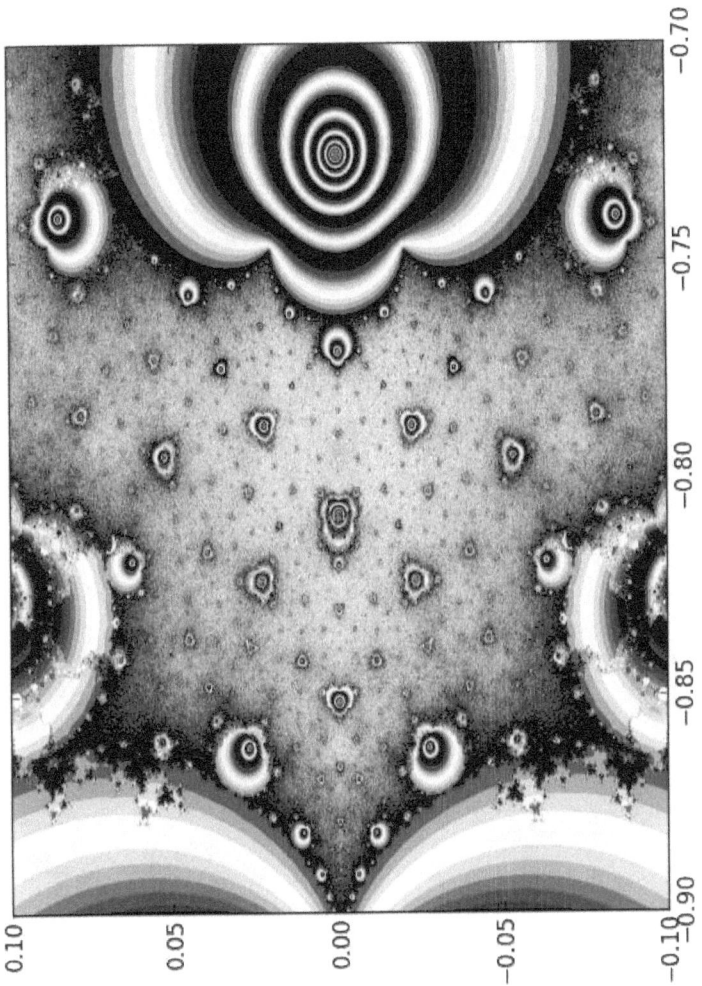

Fig. 2.7. Carpet invasion.

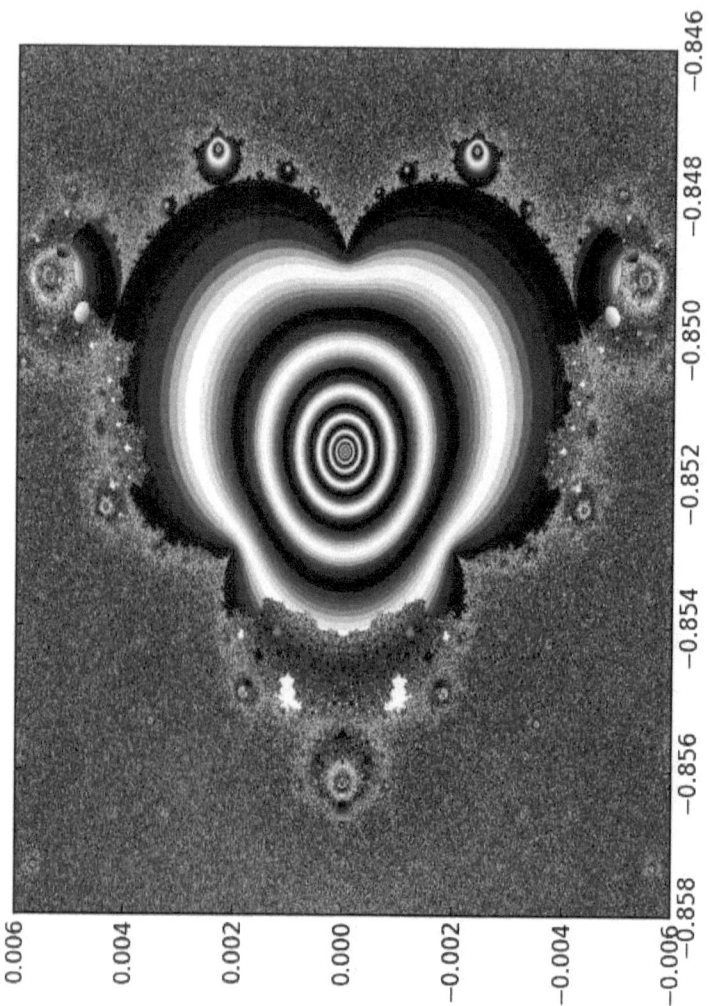

Fig. 2.8. Rust.

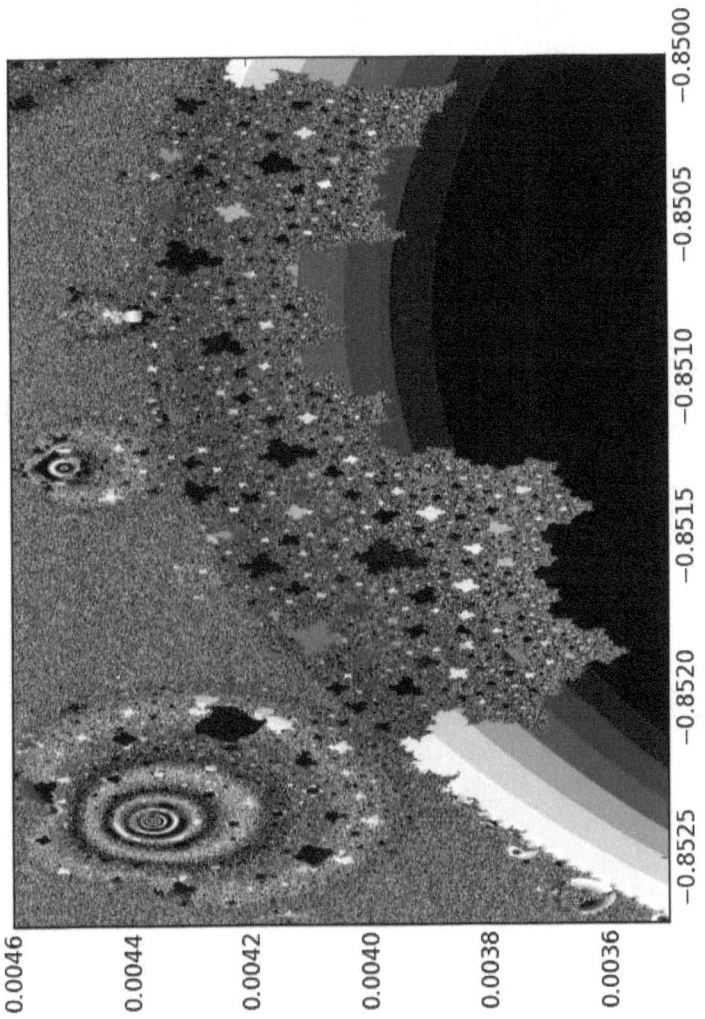

Fig. 2.9. Destruction 1.

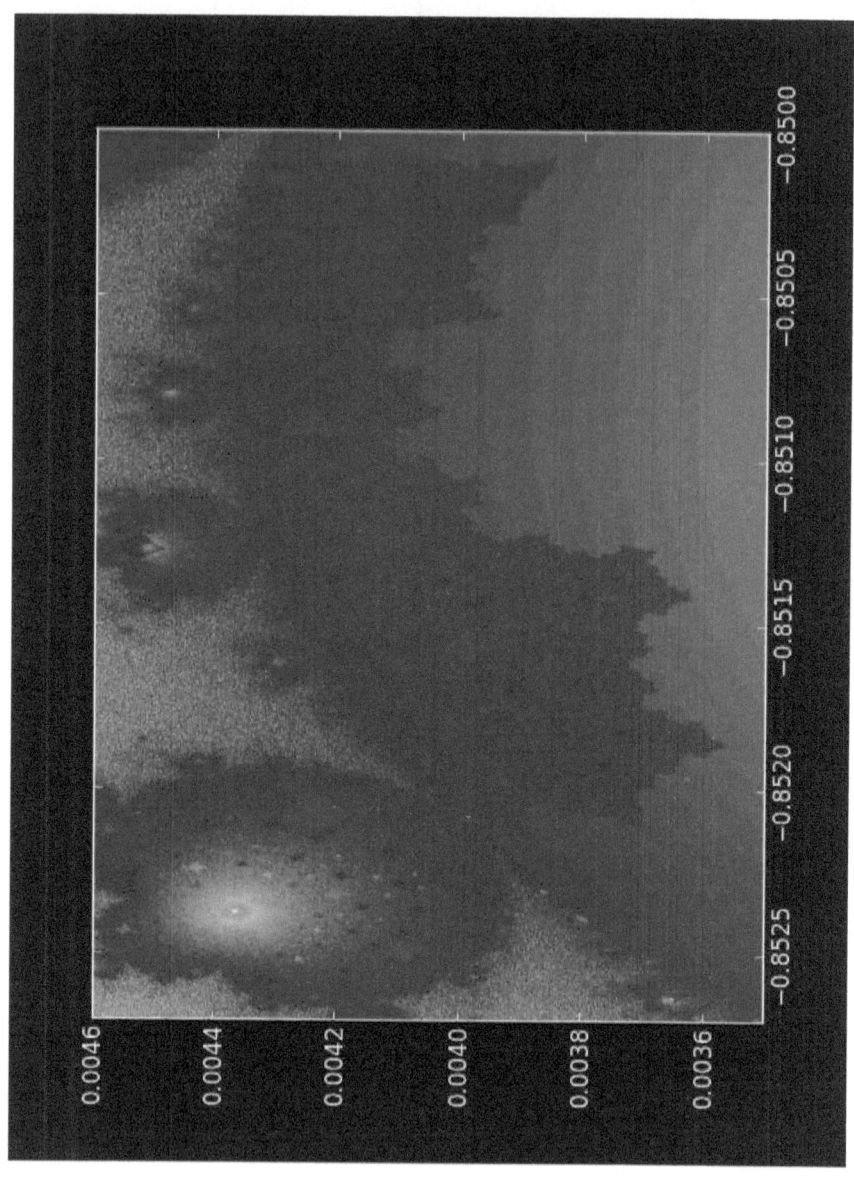

Fig. 2.10. Year 2020.

Part 3. The power complex

The fractals on Figs. 3.1 to 3.22 were generated from the following requrence:
$$z \to z^z + c \qquad (3.1)$$
where z and c are complex numbers. The shade on Figs 3.1 to 3.10 correspond to the numbers of iterations for $|z|=\infty$. The shade on Figs 3.11 to 3.20 correspond to the numbers of iterations for $|z|=2$. The shade on Figs 3.21 to 3.22 correspond to the Lyapunow exponent value, given by:

$$\lambda = \lim_{n \to \infty} \frac{1}{n} \sum_{i=1}^{n} \ln |z_i^z(1 + \ln z_i)| \qquad (3.2)$$

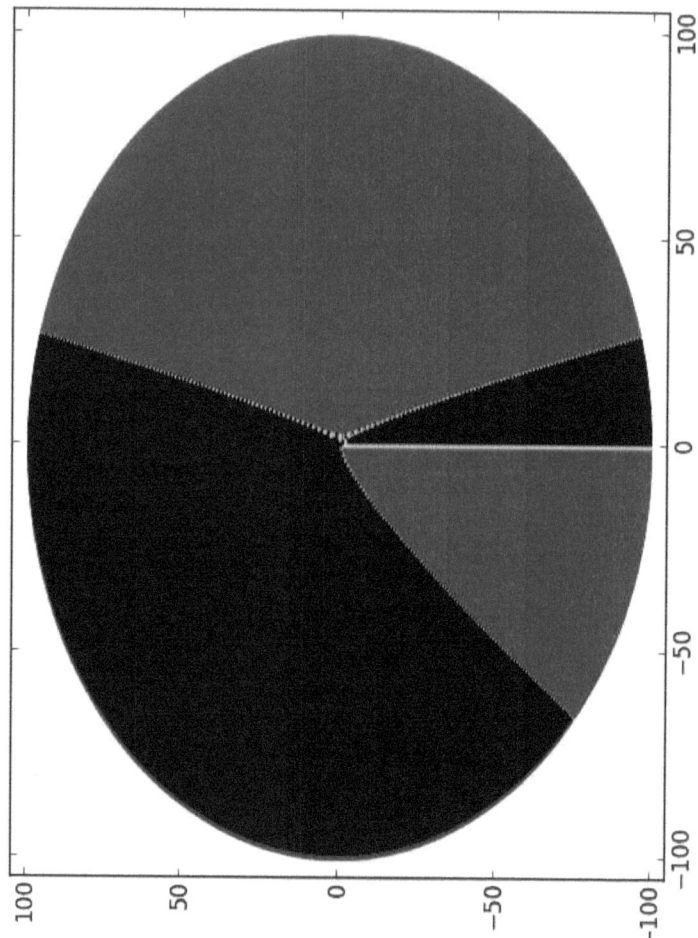

Fig. 3.1. Full set for $|z| \to \infty$.

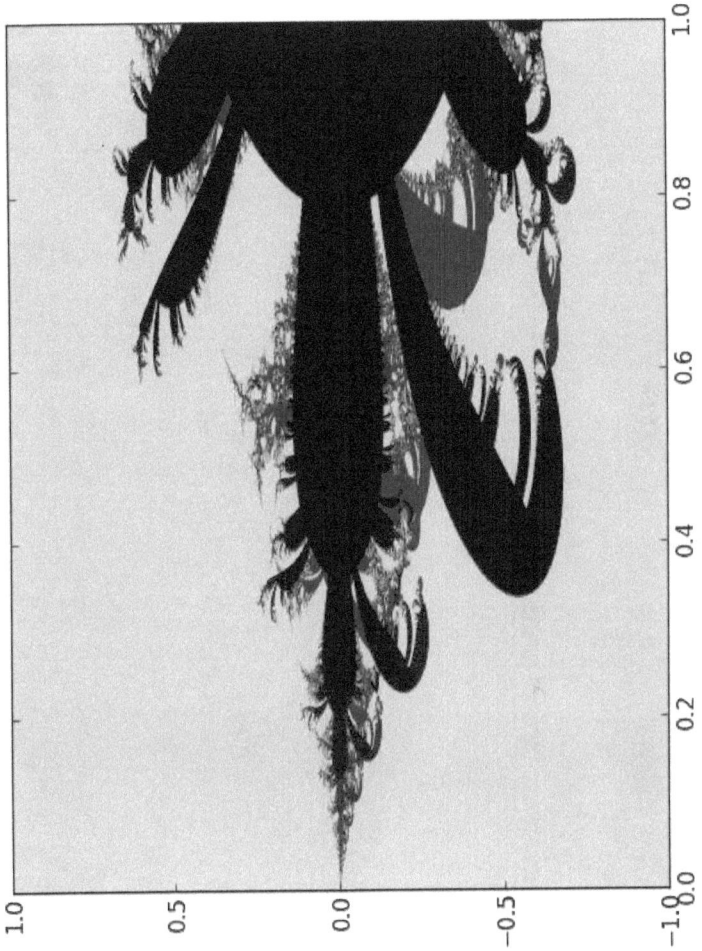

Fig. 3.2. Spear.

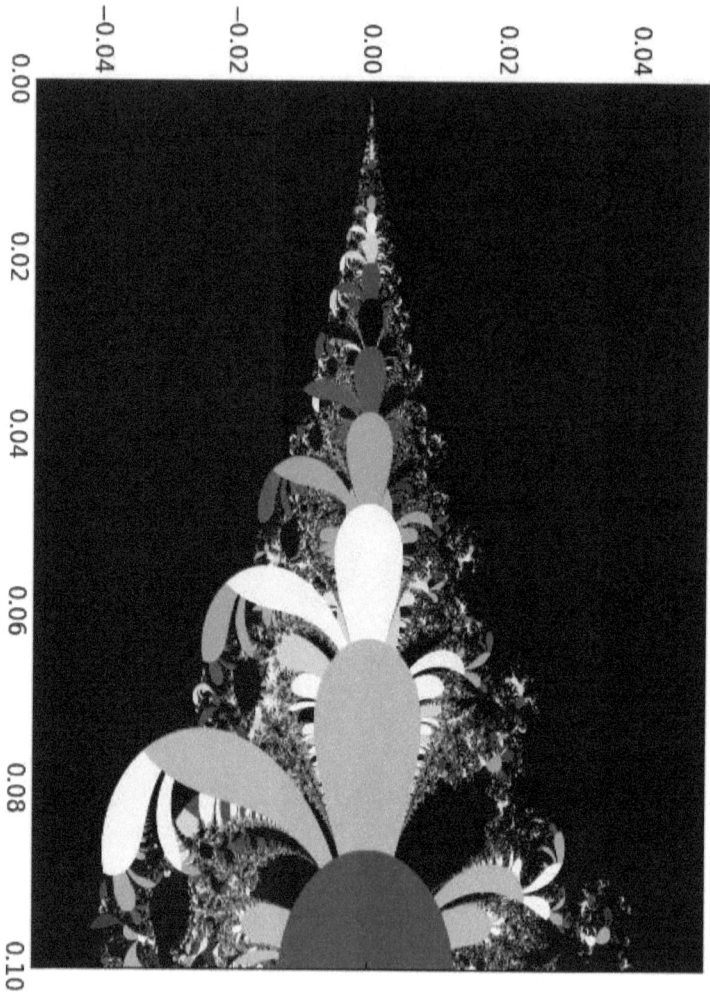

Fig. 3.3. Christmas Tree.

Fig. 3.4. Flowers.

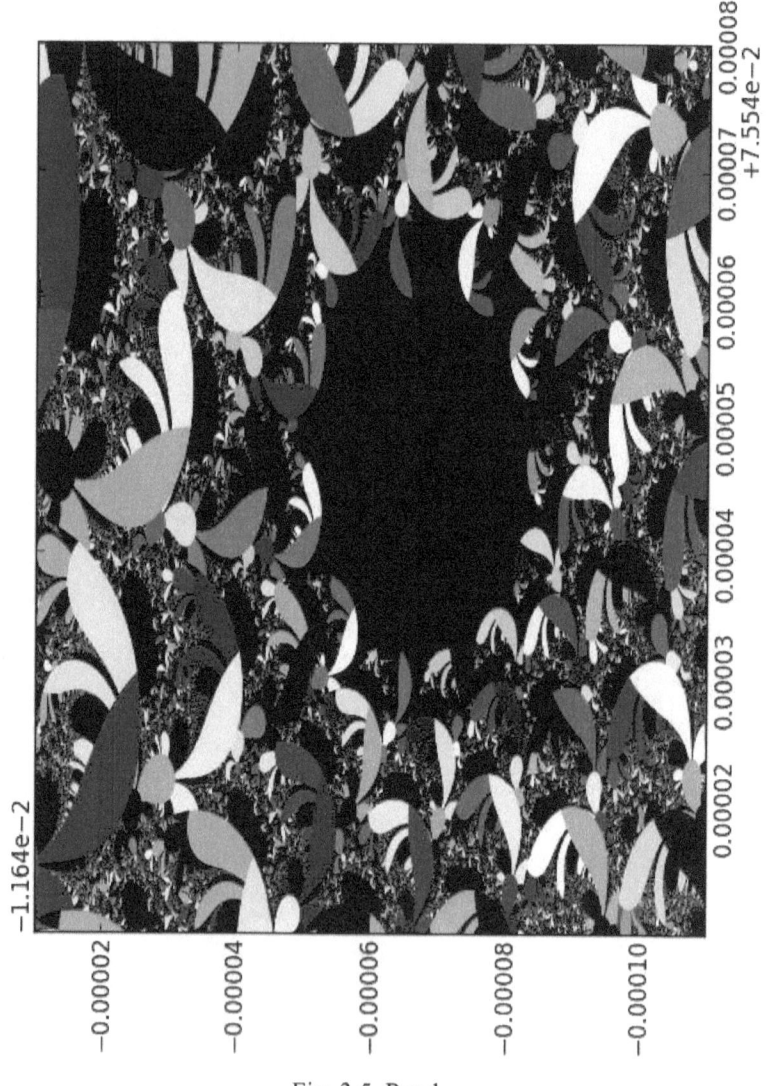

Fig. 3.5. Pond.

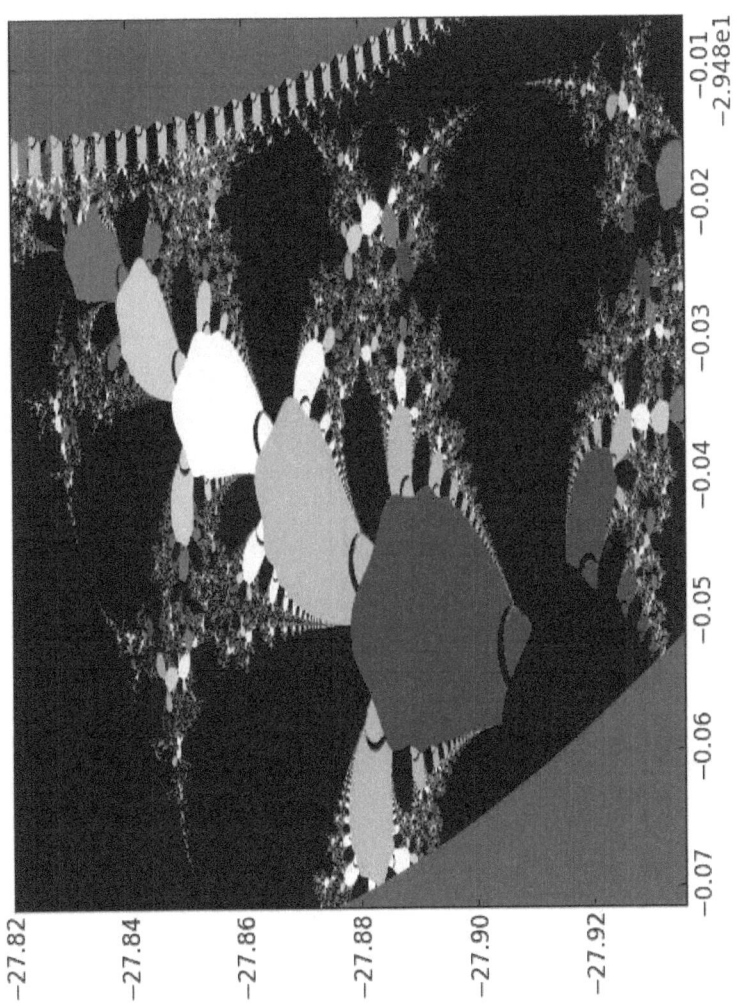

Fig. 3.6. Leaves on the water.

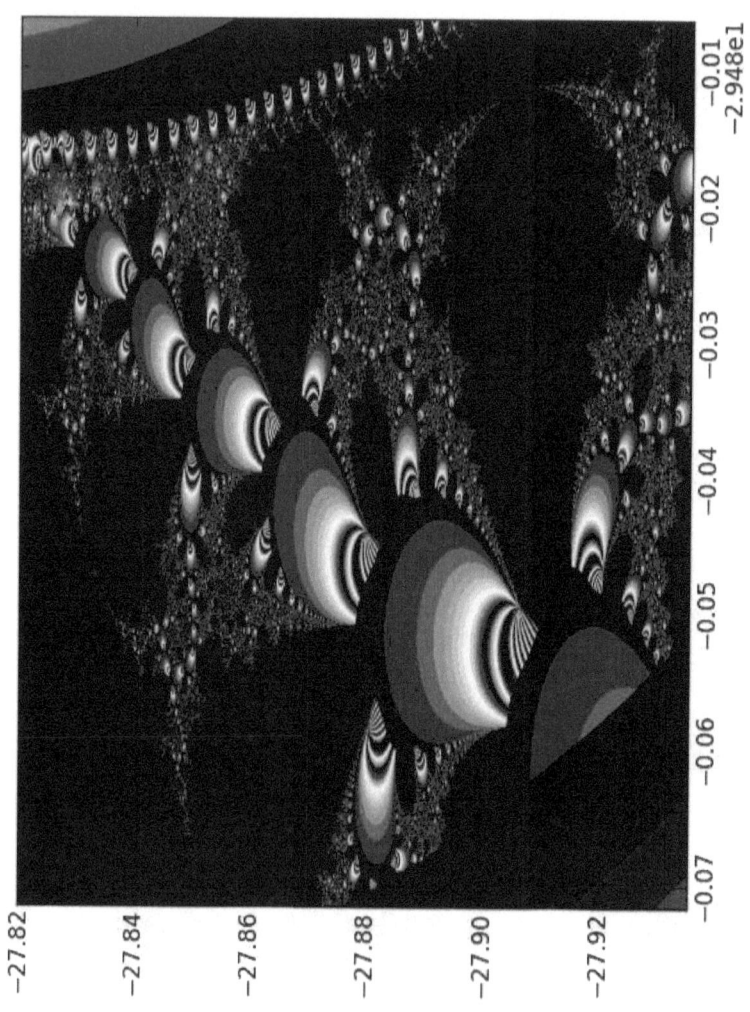

Fig. 3.7. Mussels.

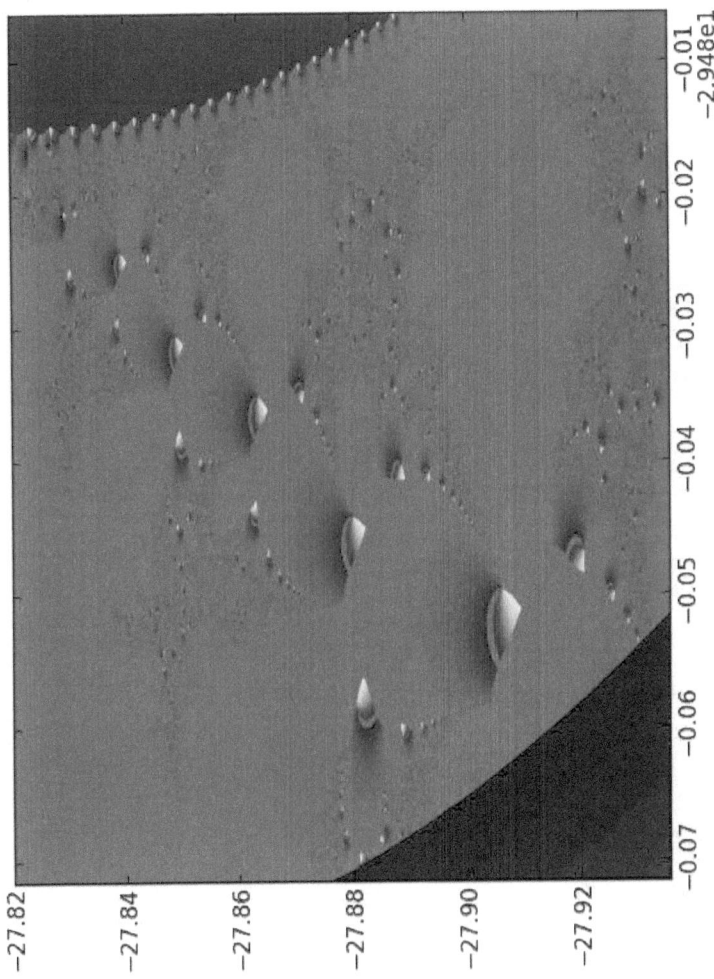

Fig. 3.8. Space odysea.

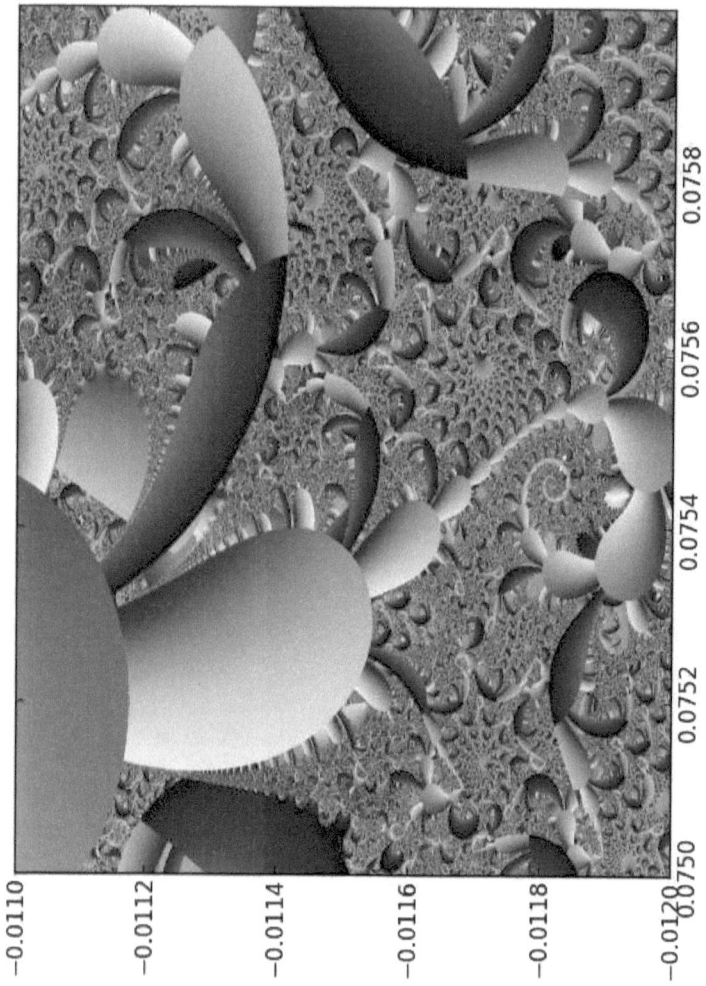

Fig. 3.9. Scorpions.

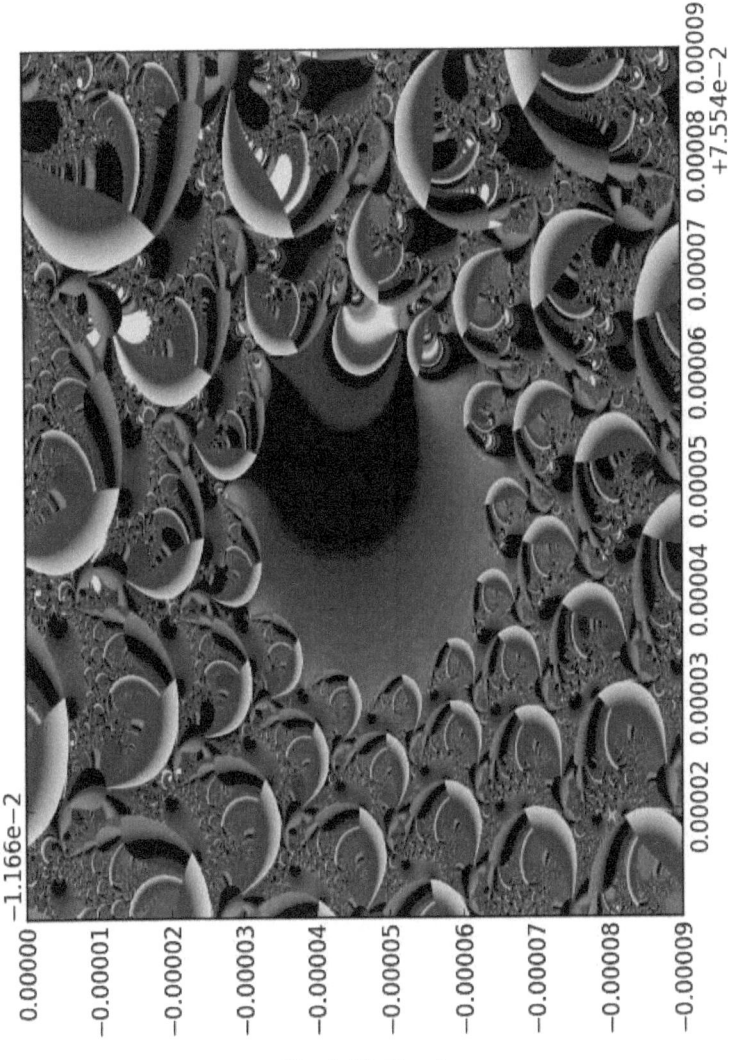

Fig. 3.10. Shoal.

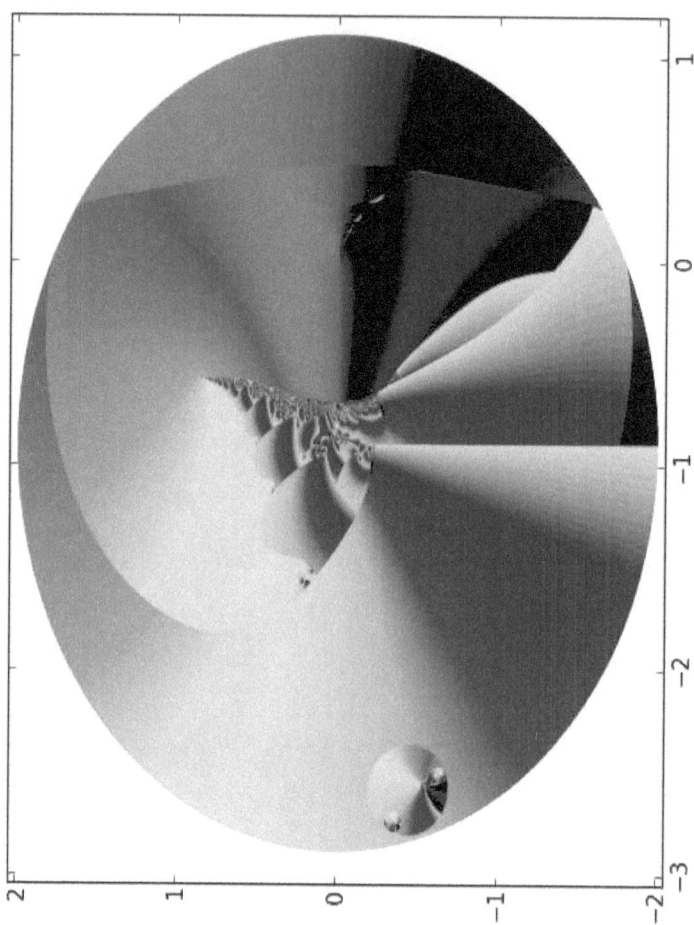

Fig. 3.11. Full set for $|z| = 2$.

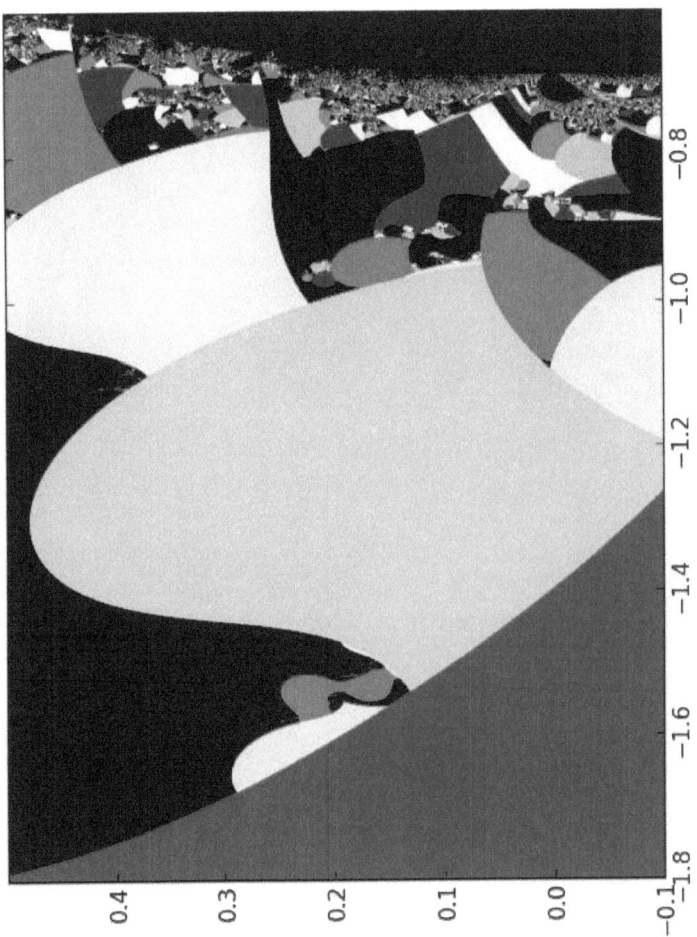

Fig. 3.12. Expedition to the mountains.

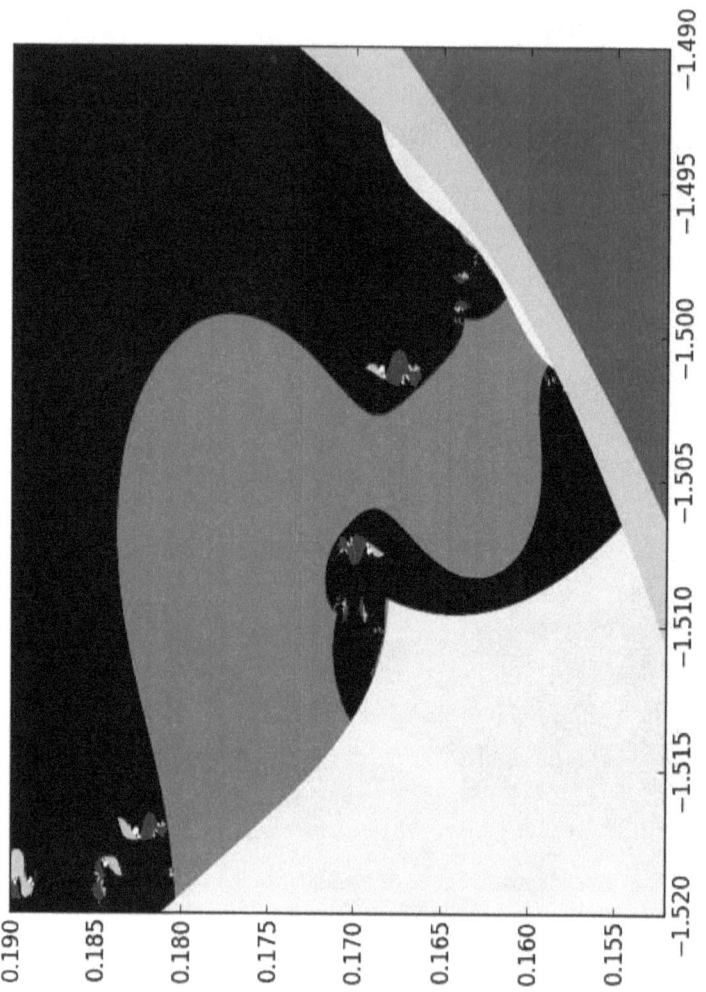

Fig. 3.13. Smurfs in the mountains.

Fig. 3.14. Bivouac in the woods.

Fig. 3.15. Destruction 2.

Fig. 3.16. Lightning.

Fig. 3.17. Clouds.

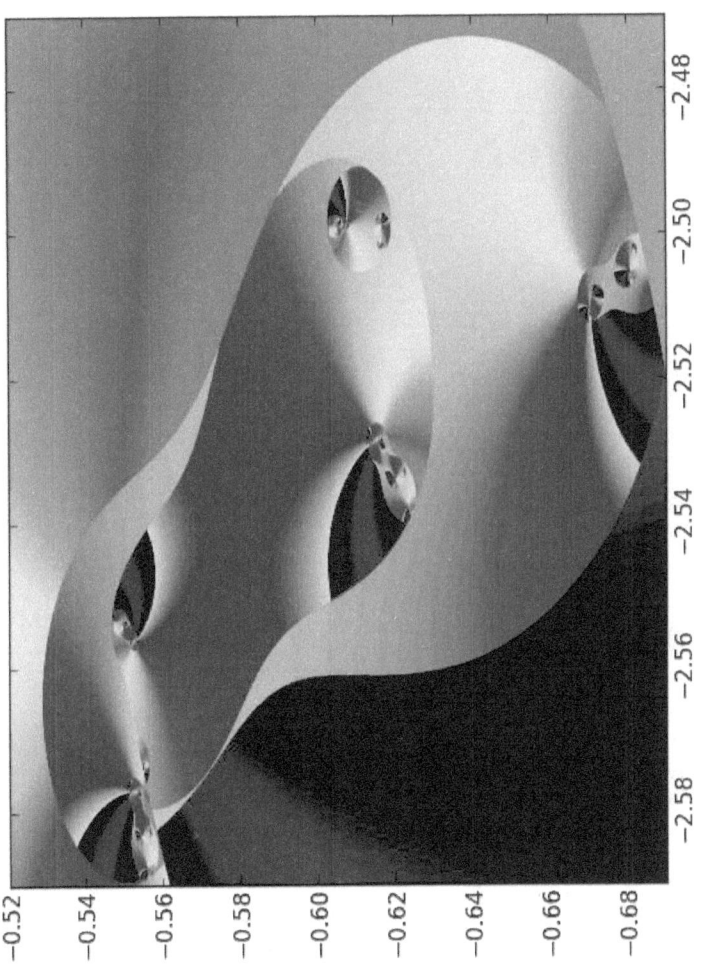

Fig. 3.19. Iron birds.

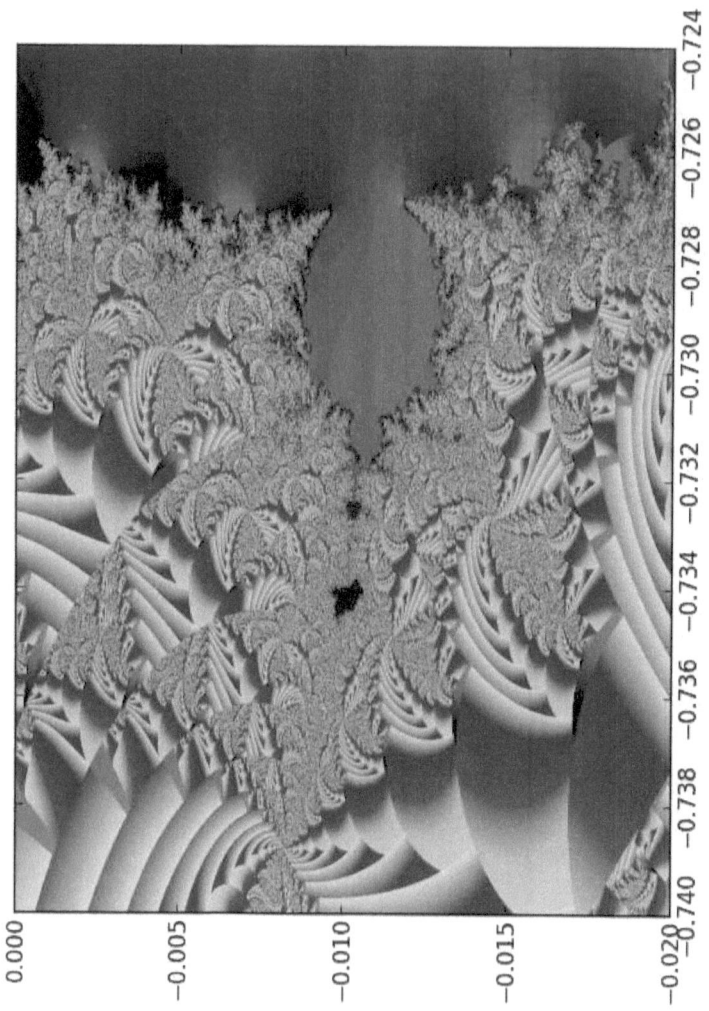

Fig. 3.19. Broch.

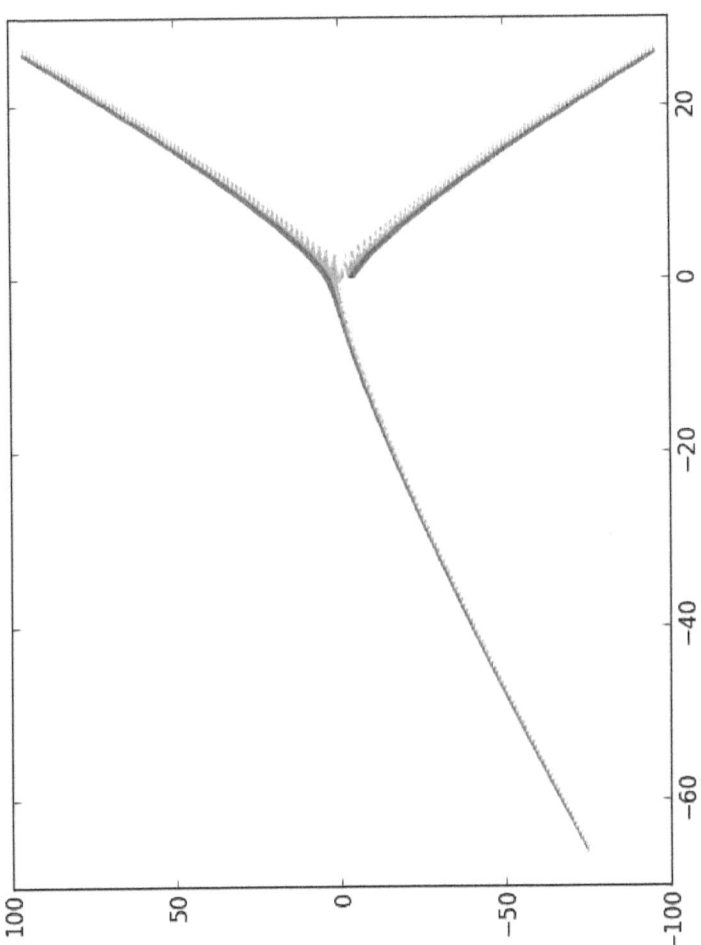

Fig. 3.20. Full set for Lyapunov exponent value.

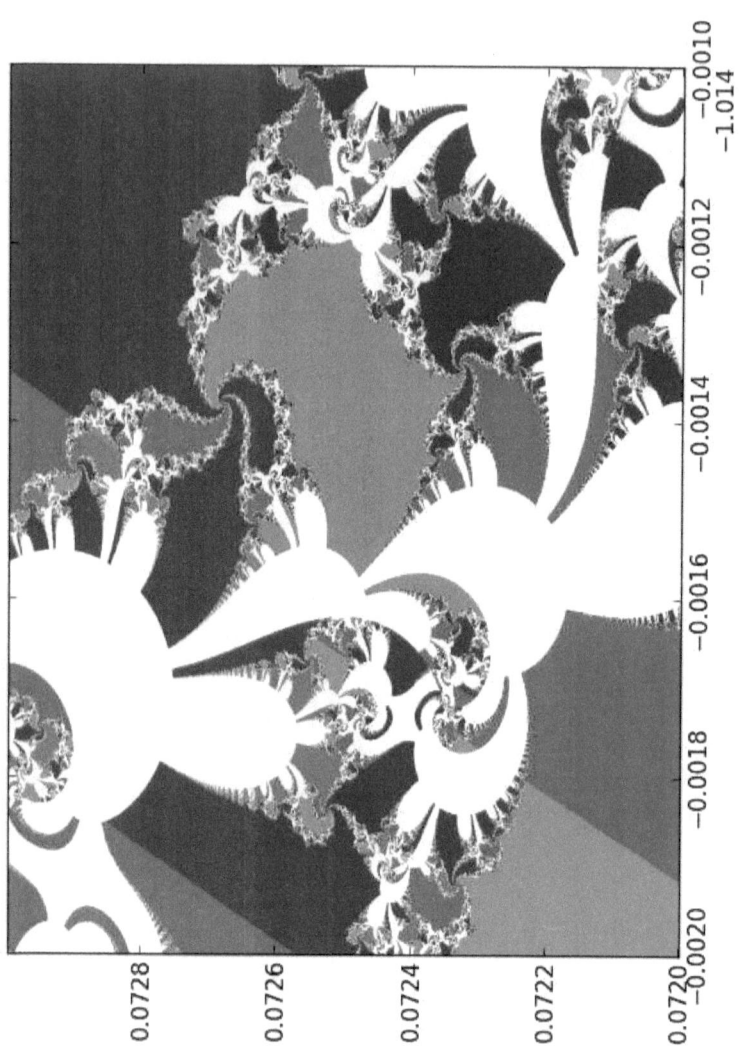

Fig. 3.21. Europe.

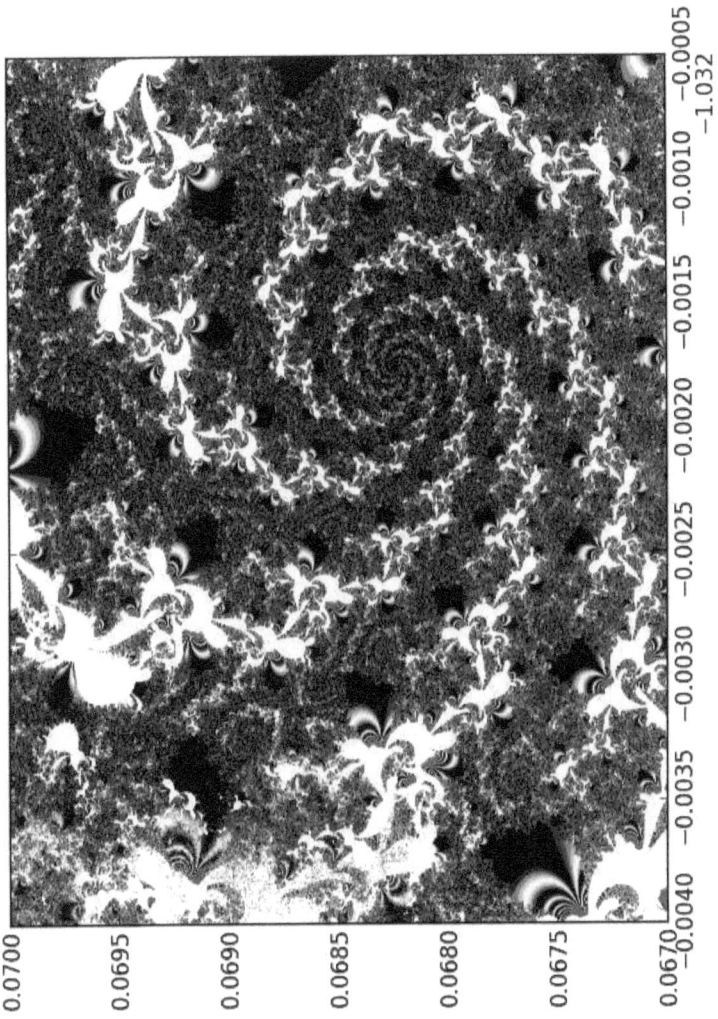

Fig. 3.22. Swirls.

Part 4. Chemical reactor models

The fractals on Figs. 4.1 to 4.12 were generated from the following chemical reactor model with complex variables [2,3]:

$$\frac{\partial \alpha}{\partial \tau} + \frac{\partial \alpha}{\partial \xi} = (1 - f_\alpha)\phi(\alpha, \Theta) \tag{4.1}$$

$$\frac{\partial \Theta}{\partial \tau} + \frac{\partial \Theta}{\partial \xi} = (1 - f_\alpha)[\phi(\alpha, \Theta) + \delta(\Theta_H - \Theta)] \tag{4.2}$$

with boundary conditions:

$$\alpha(0,\tau) = f_\alpha \alpha(1,\tau); \quad \Theta(0,\tau) = f_\Theta \Theta(1,\tau) \tag{4.3}$$

where:

$$\phi(\alpha, \Theta) = Da(1-\alpha)^n e^{\gamma \frac{\beta \Theta}{1+\beta \Theta}}. \tag{4.4}$$

Heat – integrated reactor: $f_\alpha = f_\Theta$; recycle reactor: $f_\alpha = 0$.

The following algorithm has been assumed. For fixed values $\alpha(0,0)$ the map of initial conditions in the coordinate system $\{\Theta_r(1,0), \Theta_i(1,0)\}$ was constructed. The shade of a point on the map depended on the number of iterative steps k, after which the length of the vector $|\Theta(1,k)|$ exceeded the given value of ε. This means that the computations were terminated, when

$$|\Theta(1,k)| \geq \varepsilon. \tag{4.5}$$

If the absolute value $|\Theta(1,k)|$ exceeded ε already in the first iterative step, no point was plotted on the diagram. In other words, the shades on the map correspond with the convergence rate of the numerical process. On the other hand, the points were coloured in red if the computational process was convergent, i.e. if after N steps the absolute value $|\Theta(1,N)|$ did not exceed ε. In such a way, changing successively $\Theta_r(1,0)$ and $\Theta_i(1,0)$ the fractal structures, as those on presented figures, were obtained.

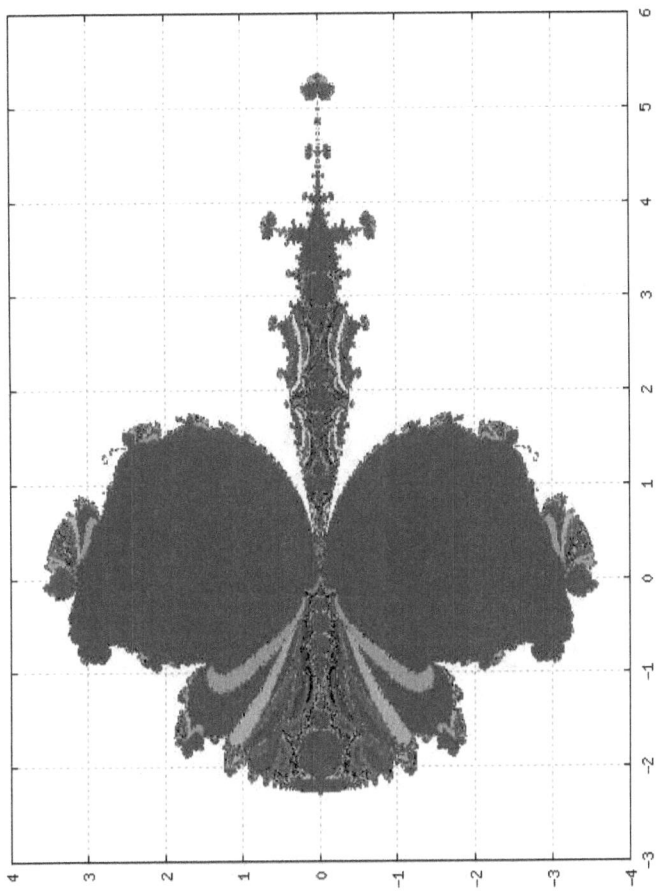

Fig. 4.1. Full set – heat integrated reactor.

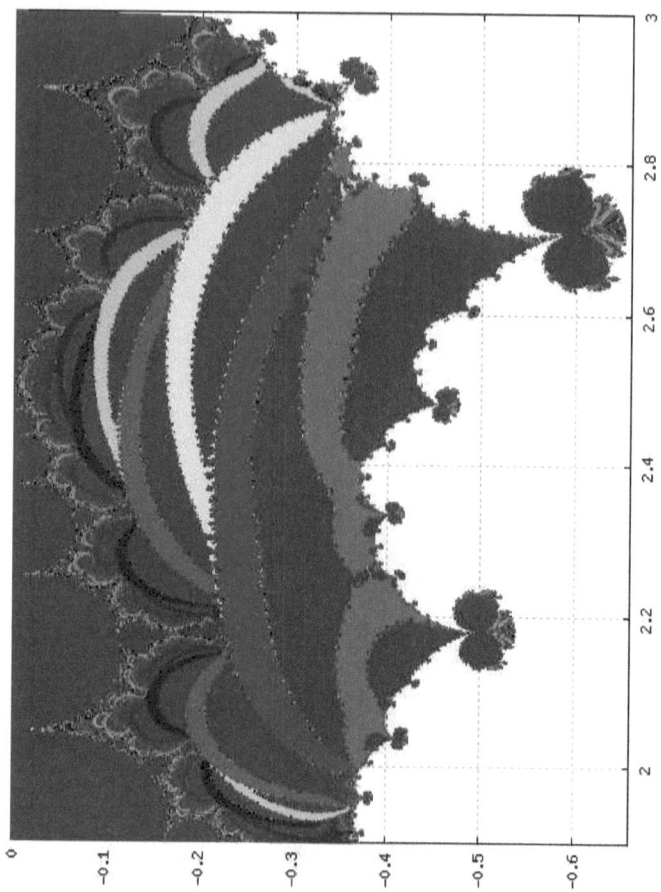

Fig. 4.2. Pageboy cap.

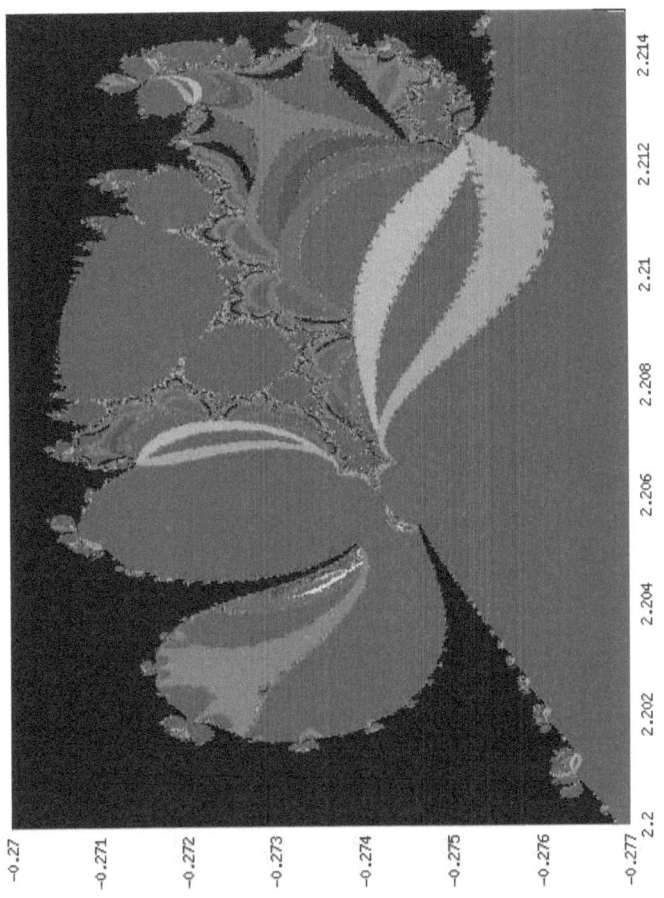

Fig. 4.3. Crown.

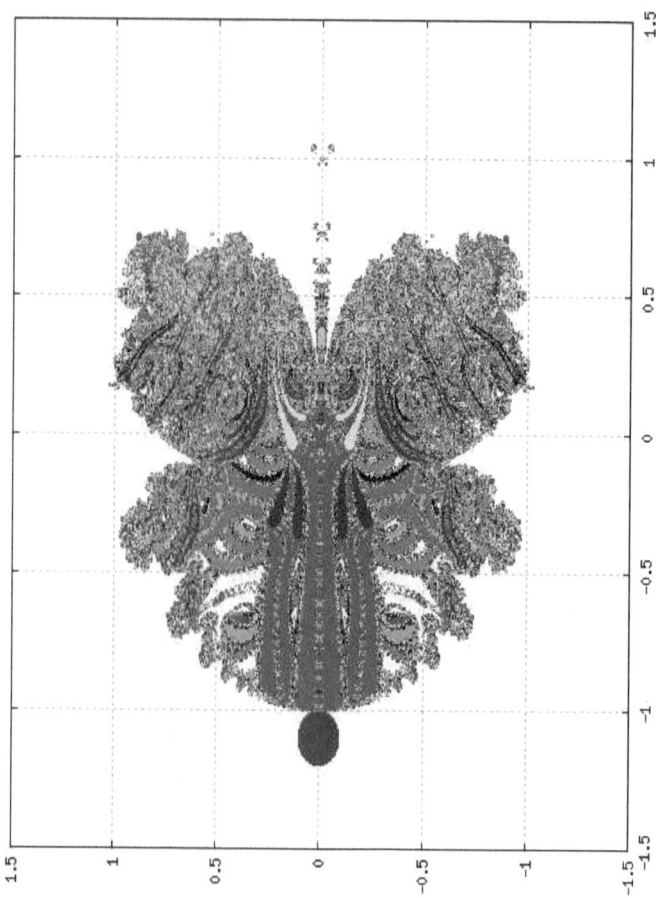

Fig. 4.4. Full set – recycle reactor.

Fig. 4.5. Hoses.

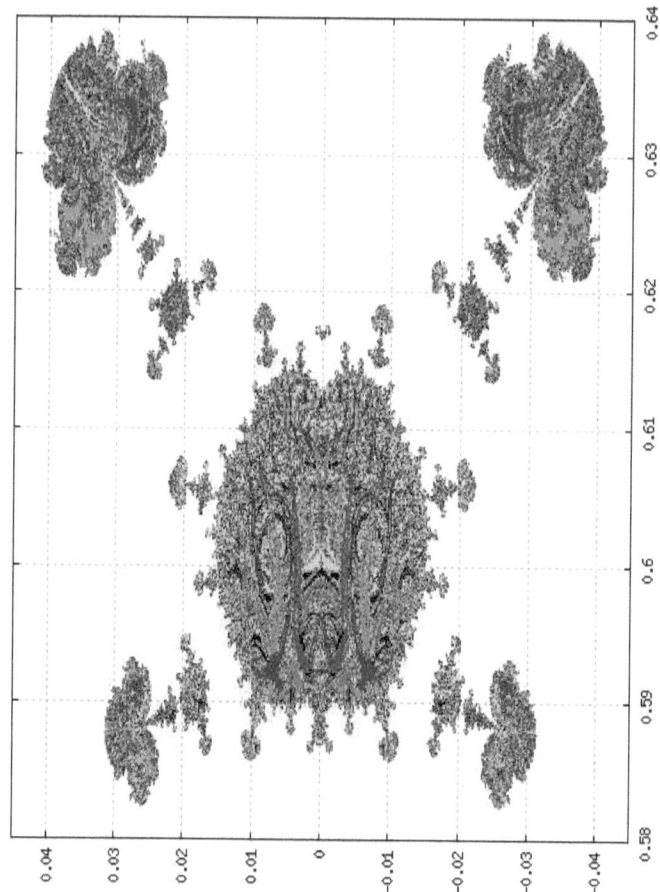

Fig. 4.6. Fleet.

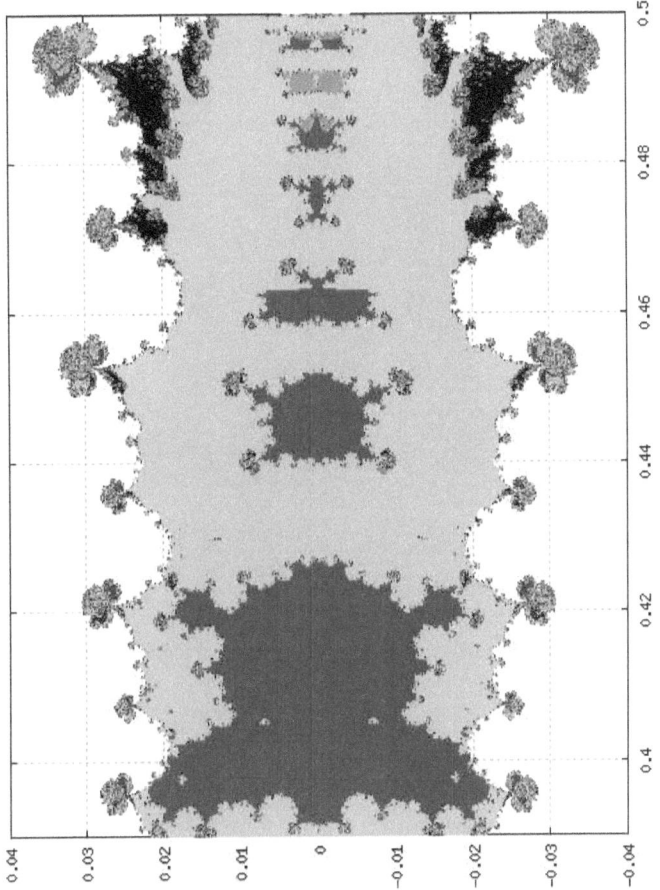

Fig. 4.7. Caterpillar.

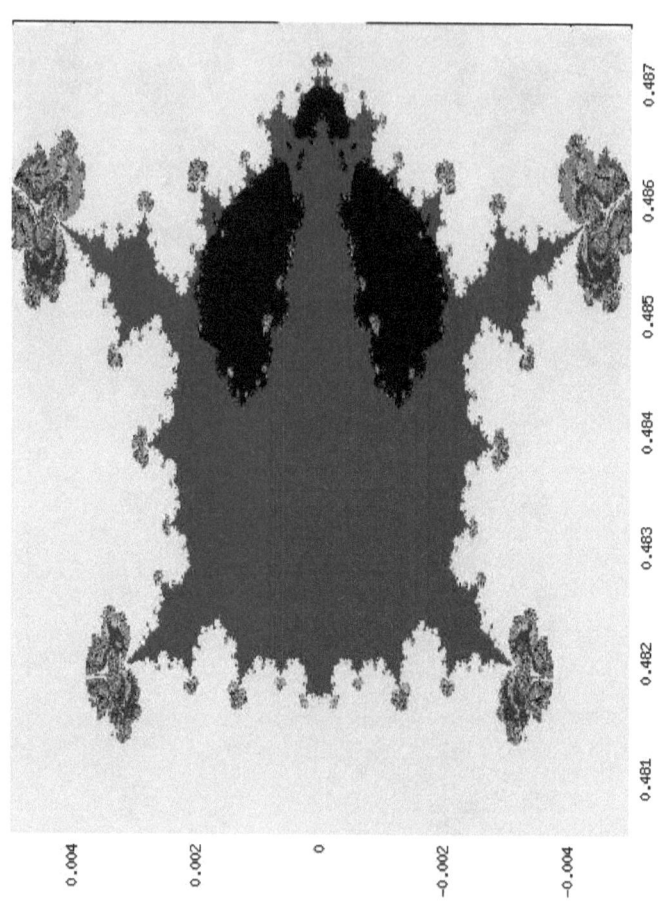

Fig. 4.8. Beetle.

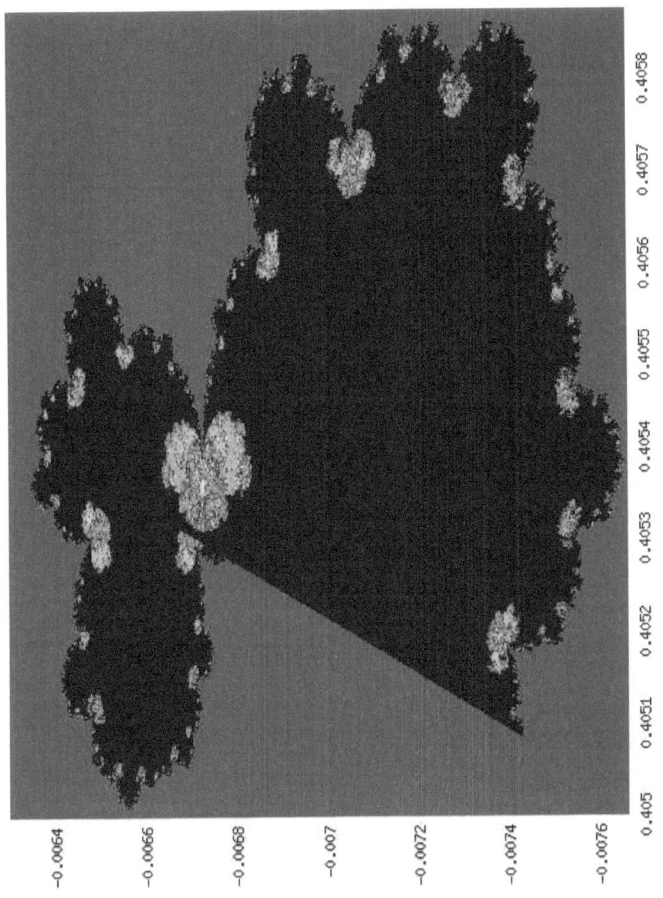

Fig. 4.9. Jewel.

Fig. 4.10. Diamonds.

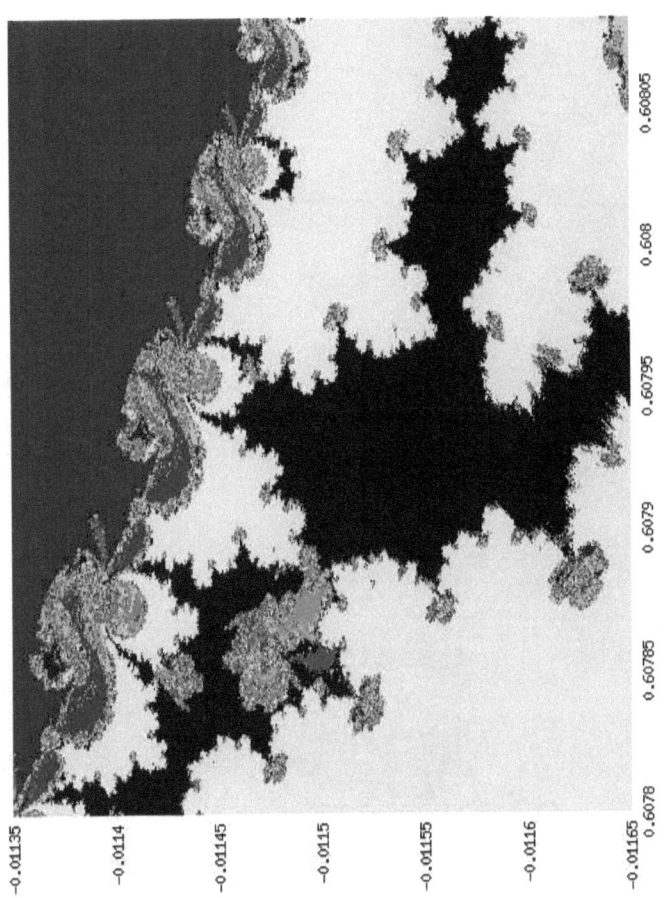

Fig. 4.11. Ocean.

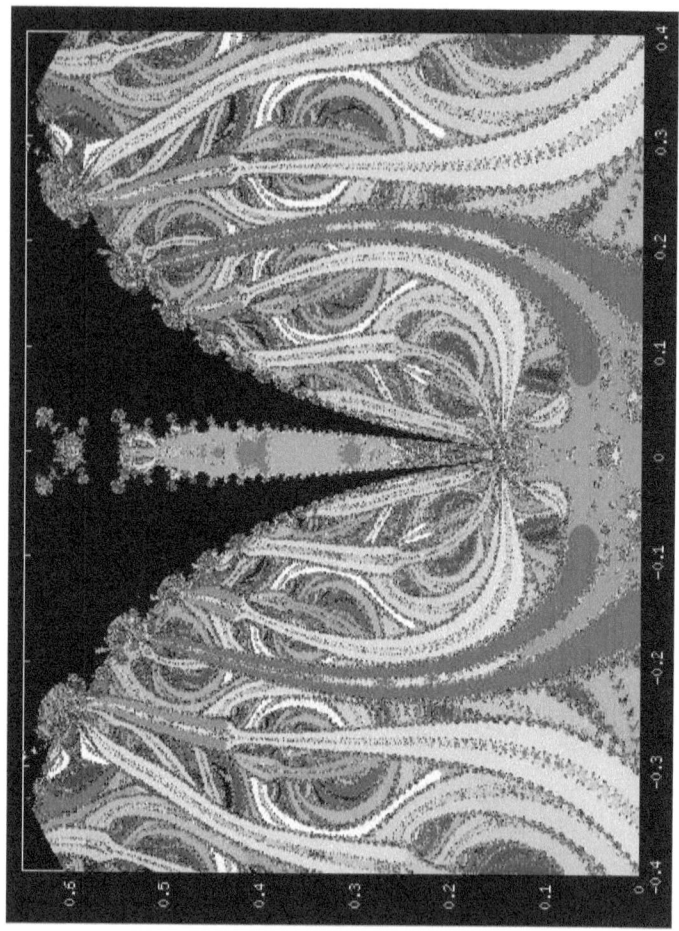

Fig. 4.12. Peacock.

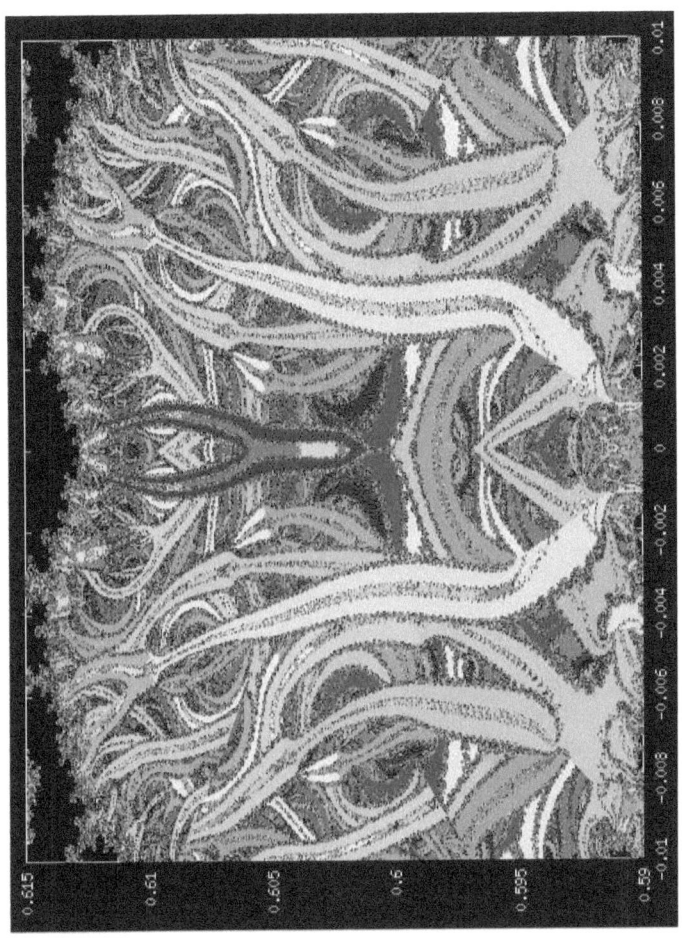

Fig. 4.13. Eden garden.

References

[1]. M. Berezowski, *Fractals gallery*: http://c504c.skroc.pl
[2]. M. Berezowski, *Fractal solutions of recirculation tubular chemical reactors*, Chaos, Solitons&Fractals, **16**, 1-12, 2003
[3]. M. Berezowski, *Fractals, bifurcations and chaos in chemical reactors*. LAP Lambert Academic Publishing, 2014, ISBN: 978-3-659-62127-7

I want morebooks!

Buy your books fast and straightforward online - at one of the world's fastest growing online book stores! Environmentally sound due to Print-on-Demand technologies.

Buy your books online at

www.get-morebooks.com

Kaufen Sie Ihre Bücher schnell und unkompliziert online – auf einer der am schnellsten wachsenden Buchhandelsplattformen weltweit! Dank Print-On-Demand umwelt- und ressourcenschonend produziert.

Bücher schneller online kaufen

www.morebooks.de

OmniScriptum Marketing DEU GmbH
Heinrich-Böcking-Str. 6-8
D - 66121 Saarbrücken

Telefax: +49 681 93 81 567-9

info@omniscriptum.de
www.omniscriptum.com

www.ingramcontent.com/pod-product-compliance
Lightning Source LLC
Chambersburg PA
CBHW031547210526
45464CB00003B/1187